Jean-Jacques Rousseau, Thomas Martyn

Thirty-Eight Plates

I0000989

Jean-Jacques Rousseau, Thomas Martyn

Thirty-Eight Plates

ISBN/EAN: 9783337375997

Printed in Europe, USA, Canada, Australia, Japan

Cover: Foto ©berggeist007 / pixelio.de

More available books at **www.hansebooks.com**

WITH

EXPLANATIONS;

INTENDED TO ILLUSTRATE

LINNÆUS's SYSTEM OF VEGETABLES,

AND PARTICULARLY ADAPTED TO THE

LETTERS ON THE ELEMENTS OF BOTANY.

———————

By *THOMAS MARTYN, B. D. F. R. & L. S. S.*

REGIUS PROFESSOR OF BOTANY

IN THE UNIVERSITY OF CAMBRIDGE.

———————

LONDON:

PRINTED FOR J. WHITE, AT HORACE'S HEAD, FLEET-STREET.

———

1799.

ADVERTISEMENT.

Sᴏᴍᴇ perfons, who have honoured the *Letters on the Elements of Botany* with their approbation, having fignified a wifh that the fubject might be ftill farther illuftrated by figures, Mr. Nᴏᴅᴅᴇʀ, an ingenious artift, has been employed for this purpofe, and has both drawn and engraved thirty-eight plates. By thefe, and the explanations which are given on the oppofite page,

the

the Author hopes that he may have met the ideas of his friends.

These Plates, with their explanations, may be confidered as an entire work: but it is prefumed that they will be much more fatisfactory when ftudied jointly with the Letters.

Six plates are given to illuftrate Rouffeau's fix letters upon the moft remarkable Natural Claffes. The reft are intended to explain the Claffes of Linnæus's Syftem in their order, except the thirty-fourth, which exhibits figures of the moft remarkable Nectaries. No general plate, explanatory of the claffical ' characters, is given; both becaufe it has already been elegantly done by Mr. Curtis, and alfo may eafily be collected from the particular plates of this work.

Thus

(v)

Thus the character of the Class

MONANDRIA is explained in — Plate VII.

DIANDRIA — — — VIII.

TRIANDRIA DIGYNIA — — IX.

———————MONOGYNIA — — X.

TETRANDRIA — — — XI.

PENTANDRIA MONOGYNIA — XII.

——————— DIGYNIA — V. and XIII.

HEXANDRIA — — I. and XIV.

HEPTANDRIA }
OCTANDRIA } — — XV.

ENNEANDRIA }
DECANDRIA } — — XVI.

DODECANDRIA — — XVII.

ICOSANDRIA — — XVIII.

POLYANDRIA — — XIX.

DIDYNAMIA — — IV. and XX.

TETRADYNAMIA — — II. and XXI.

MONADELPHIA — — XXII.

DIADELPHIA — — III. and XXIII.

POLYADELPHIA — — XXIV.

SYNGENESIA — — — VI.

———————POLYGAMIA ÆQUALIS — XXV.

————————————— SUPERFLUA XXVI.

SYNGENESIA

	PLATE
SYNGENESIA POLYGAMIA FRUSTRANEA &	} XXVII.
————————————————— NECESSARIA	
————————————————— SEGREGATA	XXVIII.
——————— MONOGAMIA —	XXIX.
GYNANDRIA — — —	XXX.
MONOECIA — — —	XXXI.
DIOECIA — — —	XXXII.
POLYGAMIA — — —	XXXIII.
CRYPTOGAMIA, FILICES — —	XXXV.
————— —————— MUSCI — —	XXXVI.
————————————— ALGÆ — —	XXXVII.
————————————— FUNGI — —	XXXVIII.

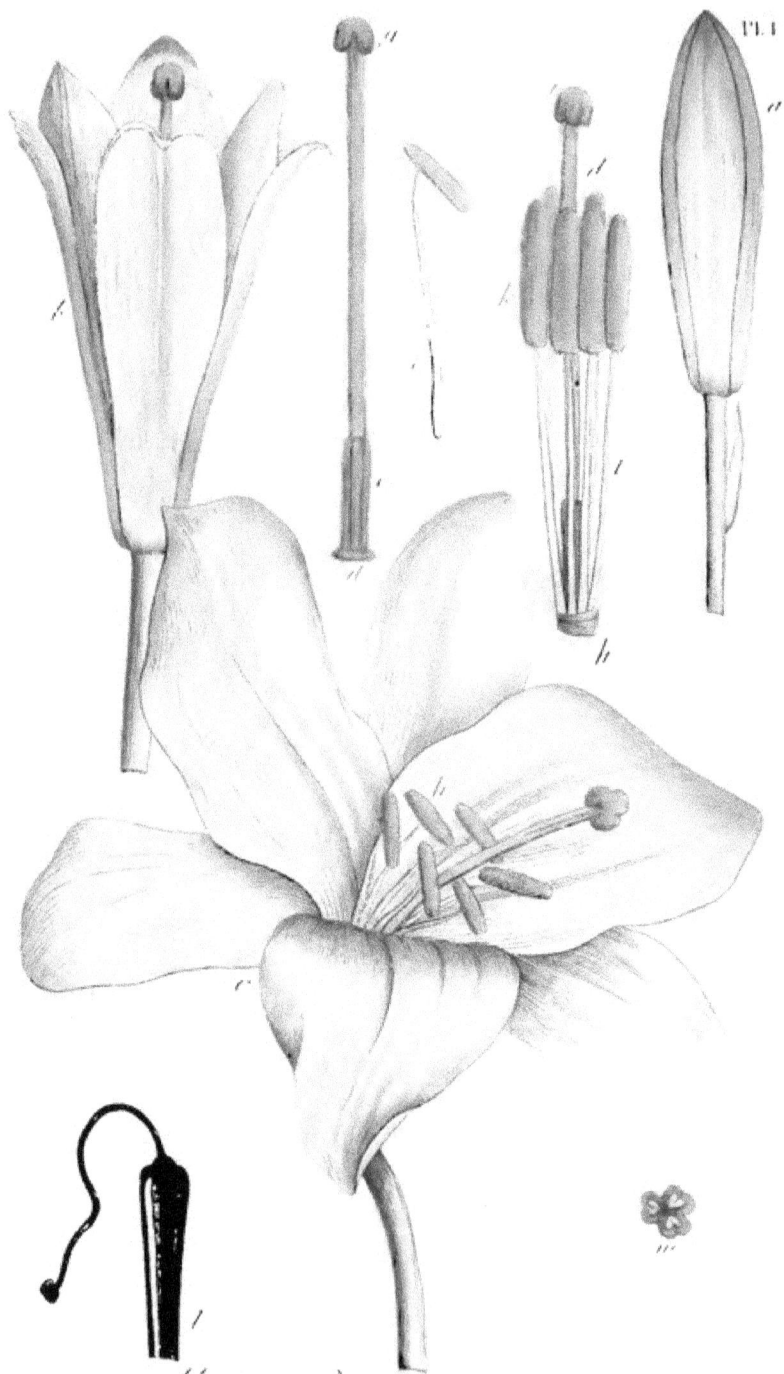

PLATE I. LETTER I.

LILIACEOUS FLOWERS.

Lilium candidum. *White Lily.*

a   The flower in bud.

b   The corolla expanding.

c   The corolla quite open.

d   The piftil or pointal.   *e* The germ.
     f The ftyle.   *g* The ftigma.

h   The fix ftamens.   *i* The filaments.
     k The anthers.

l   The germ advanced into a pericarp,
     which here is a capfule.

m   A tranfverfe fection of the pericarp, to
     fhow the three cells and feeds.

B

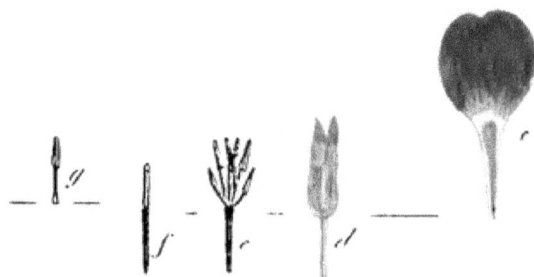

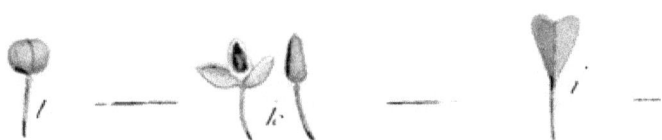

PLATE II. LETTER II.

CRUCIFORM FLOWERS.

Cheiranthus incanus. *Stock-Gilliflower*.

a A flower of the ſtock, ſhowing the four
petals and the cruciform ſhape of the
corolla.

b A back view of it, exhibiting the calyx,
conſiſting of four leaflets, and bulging
out at the bottom.

c A ſingle petal ſeparated, to ſhow the
lower narrow part, called *unguis*, or
the tail; and the upper ſpreading
part, named *lamina*, or the border,
emarginate or notched at the end.

d A ſection of the calyx, with the ſingle
piſtil and ſix ſtamens in their proper
ſituation.

e The ſix ſtamens, two of which are ſenſi-
bly ſhorter than the other four.

f The piſtil ſeparated from the other parts.

g A ſingle ſtamen.

h The fruit, ſeed-veſſel, or pericarp, called
a ſilique, opening from the bottom

upwards, and fhowing the two valves,
with the feeds ranged along the dif-
fepiment, or partition, of the two
cells, and the permanent ftigma at
the top.

i k l Figures of filicles, or fmall fhort pods
or pouches.

i The flat triangular, or heart-fhaped filicle
of the fhepherd's purfe.

k The oblong filicle of fcurvy-grafs, both
fhut and open.

l The almoft fpherical filicle of candy-tuft.
See Letter **XXIII.** *and* Plate XXI.

e Explains the claffical character of the clafs
Tetradynamia, and

h i k l Explain the characters of the two
orders, *Siliquofa* and *Siliculofa,* into
which it is divided.

Plate III

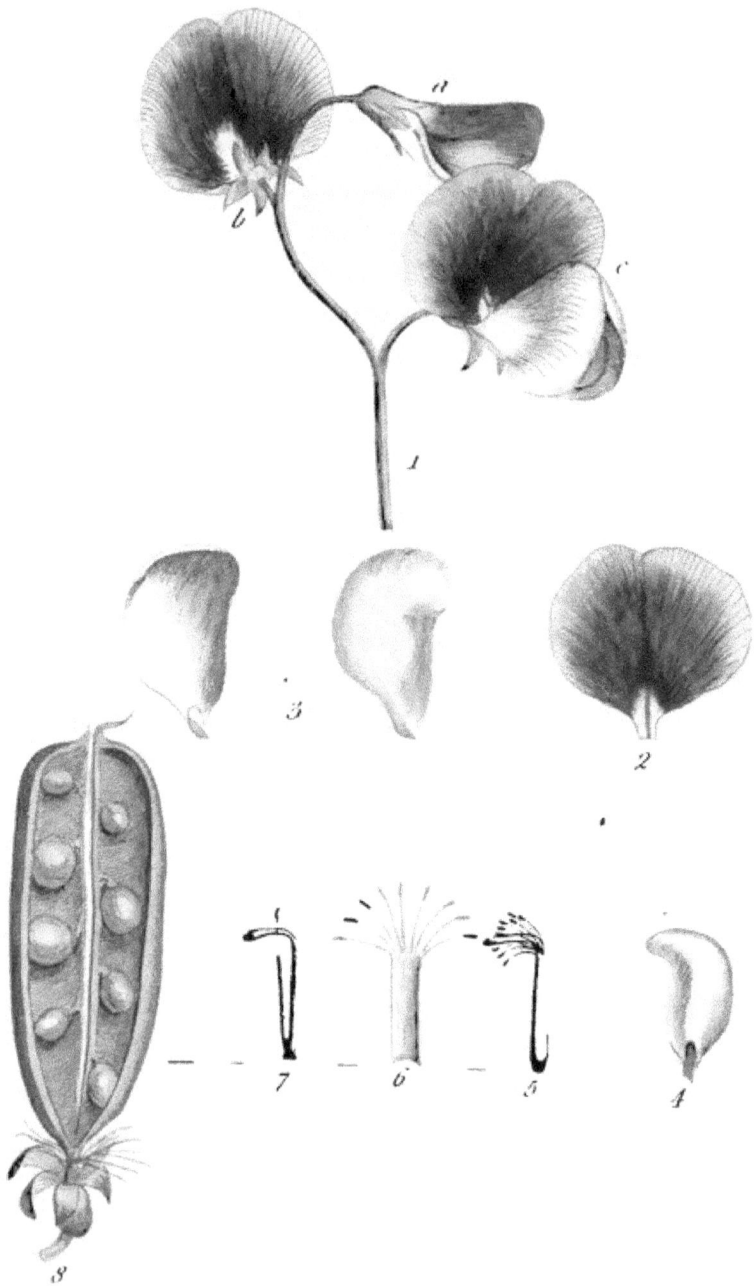

PLATE III. LETTER III.

PAPILIONACEOUS FLOWERS.

Pifum fativum.   *Garden Pea.*

Fig. 1.   The peduncle or flower-ftem of the pea, fhowing the papilio-naceous corolla in three differ-ent fituations.

a   A young flower not fully expanded.

b   An expanded flower, fhowing the back; the ftandard, or banner, fully dif-played, and the calyx cleft into five parts.

c   A fide view of an expanded flower, fhowing the banner, wings, and keel in their natural fituation.

Fig. 2.   The banner *(vexillum)*, obcordate or inverfely heart-fhaped, and emarginate.

3.   The two wings *(alæ)*.

4.   The keel *(carina)*.

5.   The piftil and ftamens in their natural fituation.

Fig. 6. The lower broad ftamen, which involves the germ, terminating in nine filaments, with an anther on each.

7. The upper narrow filament, accompanied with the piftil.

8. The pericarp, which is a legume, or pod, open to fhow the two valves and the feeds faftened alternately to the futures of the valves at the back of the legume. The permanent calyx is alfo here exhibited.

Obf. The charaƈter of the clafs *Diadelphia,* and of the order *Decandria,* as alfo of the natural clafs of *Leguminous* plants, is here explained.

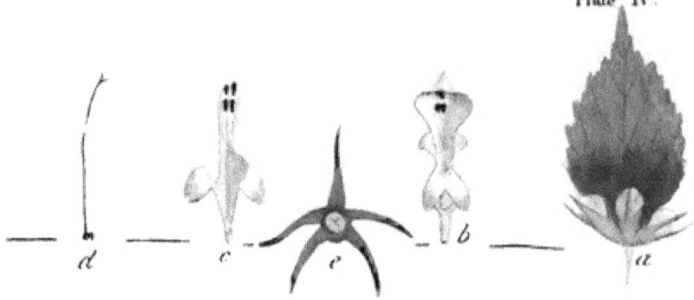

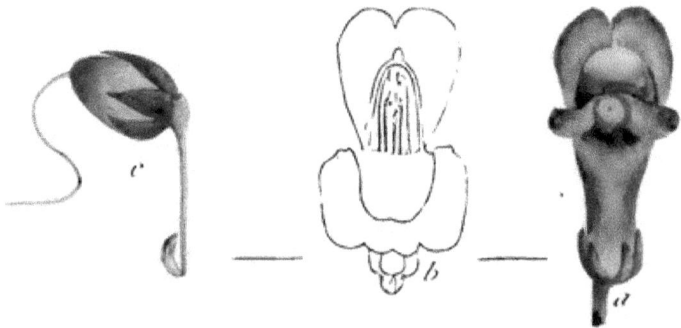

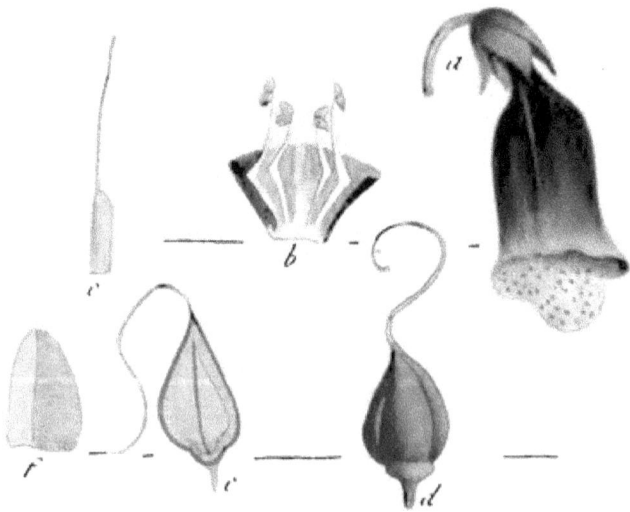

PLATE IV. LETTER IV.

RINGENT FLOWERS.

Fig. 1. Lamium album. *White Dead Nettle.*

a   Part of a whorl of flowers, showing how
they grow in the bosom of a leaf.

b   A single flower, showing the structure
of a labiate or ringent corolla, and of
that of the Lamium in particular.

c   The corolla cut away, in order to show
more distinctly the situation of the
stamens and the classical character.

d   The germs, with the style.

e   The calyx, with the four seeds within it.

Fig. 2. Antirrhinum majus. *Snapdragon.*

a   The closed ringent, or personate corolla,
in its natural form.

b   The corolla opened, to show the situation
of the stamens.

c   The capsule, with the permanent style
and calyx.

Fig. 3. Digitalis purpurea. *Purple Fox-glove.*

a   A single flower, showing the open bell-shaped corolla.

b   The inside, exhibiting the situation and structure of the stamens.

c   The germ, with the style.

d   The capsule, with the style permanent.

e   A section of the capsule.

f   A capsule, deprived in part of its outer skin, to show the interior texture of the coat.

Plate V.

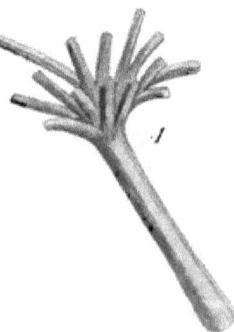

PLATE V. LETTER V.

UMBELLATE FLOWERS.

Fig. 1. Apium Petrofelinum. *Garden Parfley.*

Fig. 2. Aethufa Cynapium. *Fool's Parfley.*

a The three long leaflets of the partial involucre, fhowing a principal difference
between this and the true Parfley.

Fig. 3. Scandix Cerefolium. *Garden Chervil.*

Fig. 4. Sambucus nigra. *Common Elder.*

To fhow the difference between that
and an umbellate plant.

Fig. 5. .The flower of an umbellate plant
magnified, to fhow the parti-
cular ftructure.

Obf. Inftances of compound umbels in Fig. 1,
2, 3, and Fig. 1, 2, of Plate XIII.
A fimple umbel is reprefented at
Fig. 3, Plate XIII.

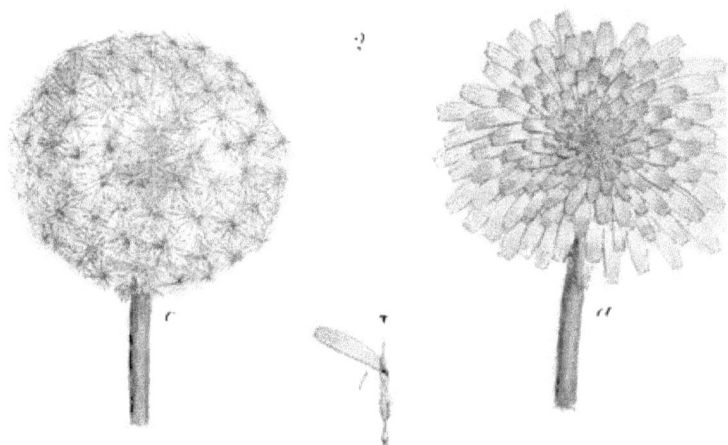

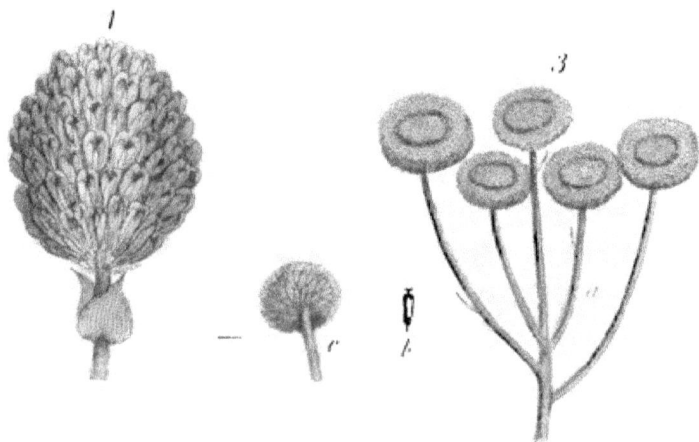

PLATE VI.   LETTER VI.

COMPOUND FLOWERS.

Fig. 1.   Bellis perennis.   *Common Daisy.*

a   The flower, which is compound and of the radiated kind, having femiflorets or ligulate florets in the ray, and tubular florets in the difk.

b   A fection of the receptacle, with the florets on it.

c   A femi-floret.

d   The cylinder of anthers, with the ftyle perforating it.

e   A floret.

Fig. 2.   Leontodon Taraxacum.   *Dandelion.*

a   The whole compound flower, confifting entirely of femi-florets, called by Linnæus ligulate florets.

b   A fingle flofcule, or floret.

c   The head of feeds.

7

Fig. 3.

 Showing a flofculous flower, or a flower compofed of florets only, called by Linnæus tubular florets.

a The whole compound flowers.

b A fingle flofcule.

c The back of a compound flower, fhow-ing the calyx.

Fig. 4. Trifolium pratenfe. *Red Clover.*

To fhow the difference between this, which is a head or aggregate of flowers, and a genuine compound flower, fuch as Fig. 1, 2, 3, exhibit.

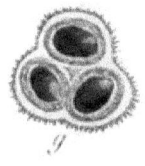

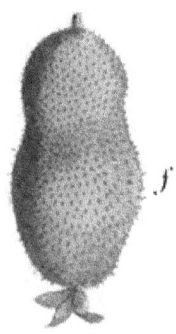

a   *c*   *b*   *a*   *a*   *d*   *c*

g   *f*   *b*   *c*   *a*   *a*

2

Drawn & engraved by Art.ᵈ Addis

PLATE VII. LETTER XI.

MONANDRIA.

Fig. 1. Canna indica. *Indian Shot.*

a a a Three different views of the flower, the corolla cut into fix lanceolated parts, one of the three interior reflected.

b The fcabrous germ, with

c The triphyllous perianth, or calyx, on the top of it.

d The anther growing to one of the petals, which ferves it for a filament.

e The ftyle, growing to the petaliform filament.

f The fcabrous capfule.

g Cut open to fhow the three cells.

Fig. 2. Hippuris vulgaris. *Mare's Tail.*

a a The germ.

b The ftamen.

c The ftyle.

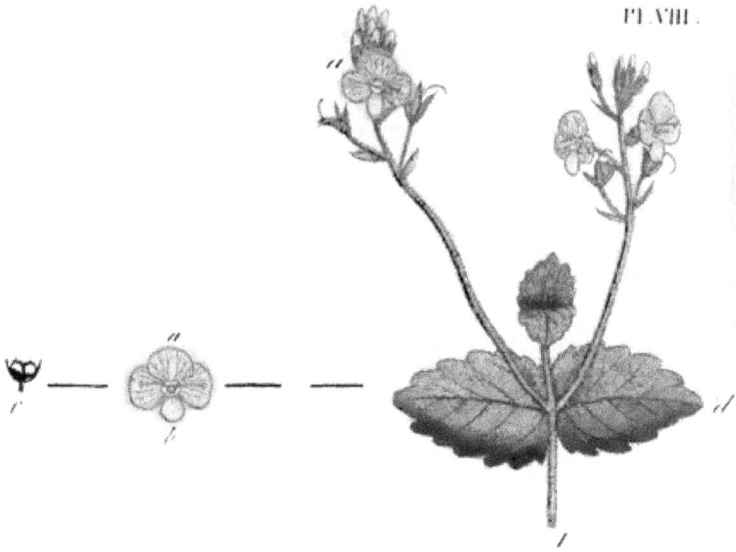

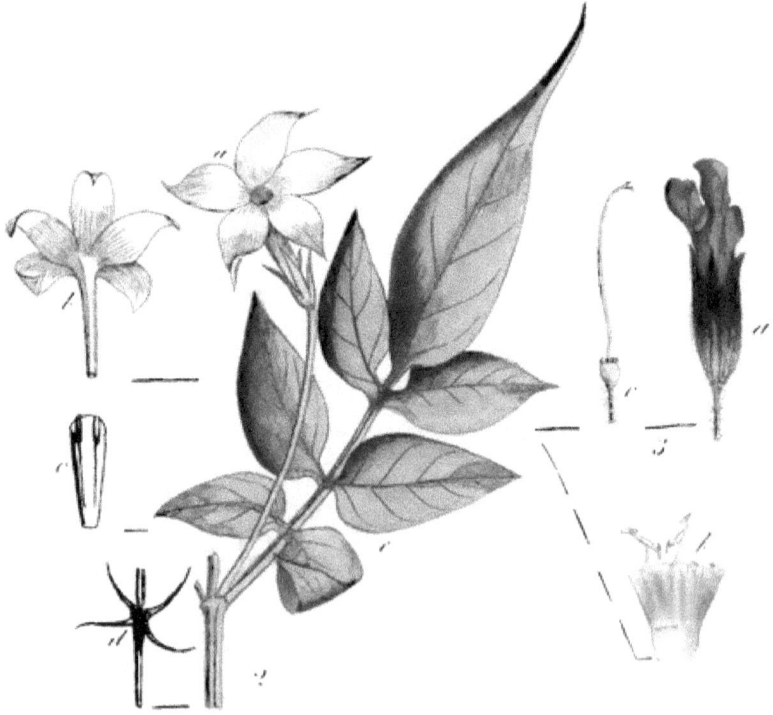

PLATE VIII. LETTER XII.

DIANDRIA.

Fig. 1.   Veronica Chamædrys. *Wild Speed-*
well.

a   The wheel-fhaped corolla, divided into
four fegments, the loweft *(b)* narrower than the reft.

c   The capfule.

d   The oval, wrinkled leaves, indented about
the edge.

Fig. 2.   Jafminum officinale.   *White Jaf-*
mine.

a   A front view of the monopetalous falver-
fhaped corolla, divided into five fegments.

b   A back view of the corolla.

c   The tube of the corolla, with the anthers
lying within it.

d   The calyx, with the rudiment of the
fruit.

e   A leaf pinnated, with all the lobes diftinct.

Fig. 3. Salvia officinalis. *Garden Sage.*

a A flower.

b The two stamens, showing their singu-
lar structure.

c The pistil separate.

2

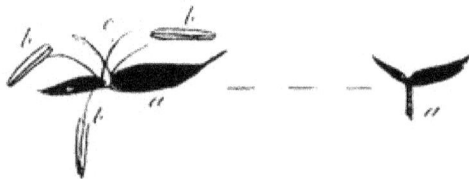

PLATE IX. LETTER XIII.

TRIANDRIA, DIGYNIA GRASSES.

Fig. 1. Lolium perenne. *Ray Grafs.*

As an inftance of a fpiked grafs.

Fig. 2. Dactylis glomerata. *Hard Grafs.*

a The chaff or glume.

b b b The three ftamens.

c The two reflected ftyles, with the feathered ftigmas.

PLATE X. LETTER XIV.

TRIANDRIA MONOGYNIA.

Iris pumila.

a   The sheath, or spathe.

b   The corolla, consisting of six parts, united at the base.

c c   The outer petals, called *falls.*

d d   The inner petals, called *standards.*

e e   The petal-form stigma, each part concealing one stamen under it.

f   A single stamen.

g   The germ, inferior or below the corolla.

h h   The nectary, in a villous line along the reflected petals.

.

Plate XI.

2

3

PLATE XI. LETTER XV.

TETRANDRIA.

Fig. 1. Scabiofa columbaria. *Small Scabious.*

An aggregate flower, confifting of many flofcules.

b A fingle flofcule ; the corolla cut into five
irregular fegments, and the germ
crowned with hairs.

c The calyx, with the four ftamens and
the piftil.

Fig. 2. Rubia peregrina. *Wild Madder.*

An inftance of ftellated plants.

The fquare ftalk : the ftellated leaves : the
corolla of four fegments : the double
germ below the flower.

Fig. 3. Plantago lanceolata. *Ribwort Plan-
tain.*

a The flowers growing in a fpike or oblong
head.

C 3

f The angular fcape.

o A fingle flower, exhibiting the quadrifid corolla and the very long filaments.

d The germ and ftyle.

e The calyx, inclofing the capfule.

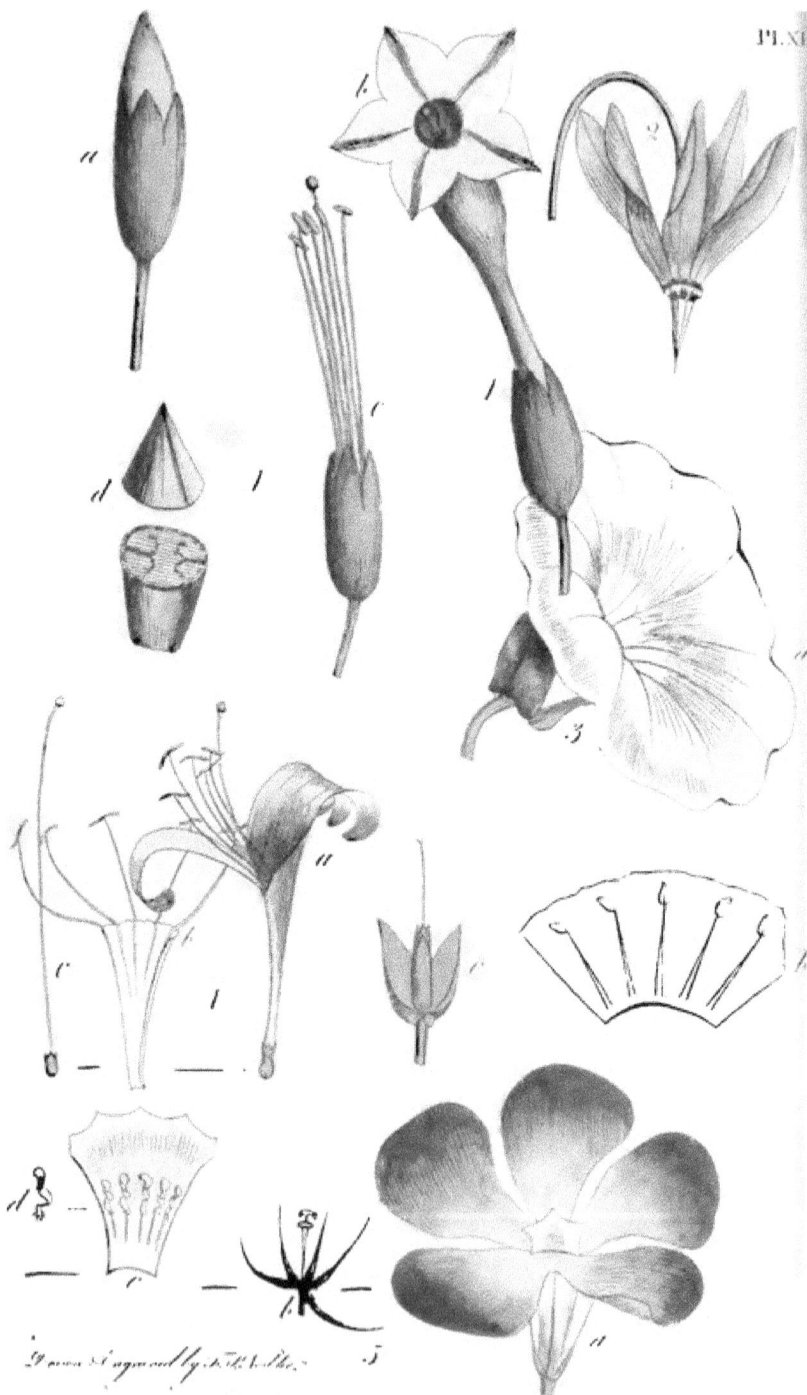

Published as May 1811 the Act directs by P.White & Son.

PLATE XII. LETTER XVI.

PENTANDRIA MONOGYNIA.

Fig. 1. Nicotiana Tabacum. *Common To-bacco.*

a A flower-bud.

b A flower, fhowing the funnel-fhaped corolla difplayed.

c The corolla removed, to fhow the five ftamens and piftil.

d A tranfverfe fection of the capfule.

Fig. 2. A flower of Dodecatheon Meadia.

Fig. 3. Convolvulus fepium. *Great Bind-Weed.*

a The corolla, with the involucre immediately below it, at Fig. 3.

b The five ftamens difplayed.

c The germ within the calyx, with the ftyle, terminated by the two ftigmas.

C 4

Fig. 4.　Lonicera Caprifolium.　*Garden Honeyfuckle.*

a　A flower, exhibiting the irregular monopetalous corolla.

b　The tube opened, to fhow the manner in
which the filaments are fixed.

c　The piftil.

Fig. 5.　Vinco major.　*Great Periwinkle.*

a　The corolla, fhowing the bending of its
five divifions, and the pentagon form
of the *faux*, or opening of the tube.

b　The calyx divided to the bottom into five
fegments; and the piftil with two
ftigmas, one over the other.

c　The tube of the corolla opened, to fhow
the fituation of the five ftamens and
form of the anthers.

d　A fingle ftamen feparate.

Pl. XIII

Published as May 1795 as the Act directs by B. White & Son

PLATE XIII. LETTER XVII.

PENTANDRIA DIGYNIA.

Fig. 1. Sium nodiflorum. *Creeping Water Parfnep.*

To fhow the difference between this plant and water creffes, reprefented in Plate **XXI**.

a A pinnated leaf, the pinnæ, fmall or com-
ponent leaves, longer and narrower
than thofe of water creffes, ferrated
on the edges and pointed at the end :
the terminating pinna trifid.

b A feffile umbel of flowers.

c A fingle flower.—*d* The fruit.

Fig. 2. Scandix Anthrifcus. *Hemlock Chervil.*

To fhow the difference between that and
Garden Chervil. Plate 5, Fig. 3.

a An umbel of flowers.

b An umbel of fruits.

Fig. 3. Scandix Pecten. *Shepherd's Needle,* or Venus's Comb.

a The umbels, being inftances of a fimple umbel.

b The feeds, terminated by the long proceffes or beaks, which gave occafion to the names.

PLATE XIV. LETTER XVIII.

HEXANDRIA.

Fig. 1. Tradefcantia Virginica. *Virginian Spiderwort.*

a   The corolla of three petals.
b b   The three-leaved calyx.
c   One of the fringed filaments.
d   The piftil.

Fig. 2. Narciffus Tazetta. *Polyanthus Narciffus.*

a   The corolla in front, fhowing the fix equal petals, and the funnel or cup-fhaped nectary.
b   A back view of the flower, fhowing that the corolla is fuperior, or on the top of the germ.
c   The fpathe.
d   The corolla opened, to fhow the fituation of the fix ftamens within the nectary.
e   The piftil.   9

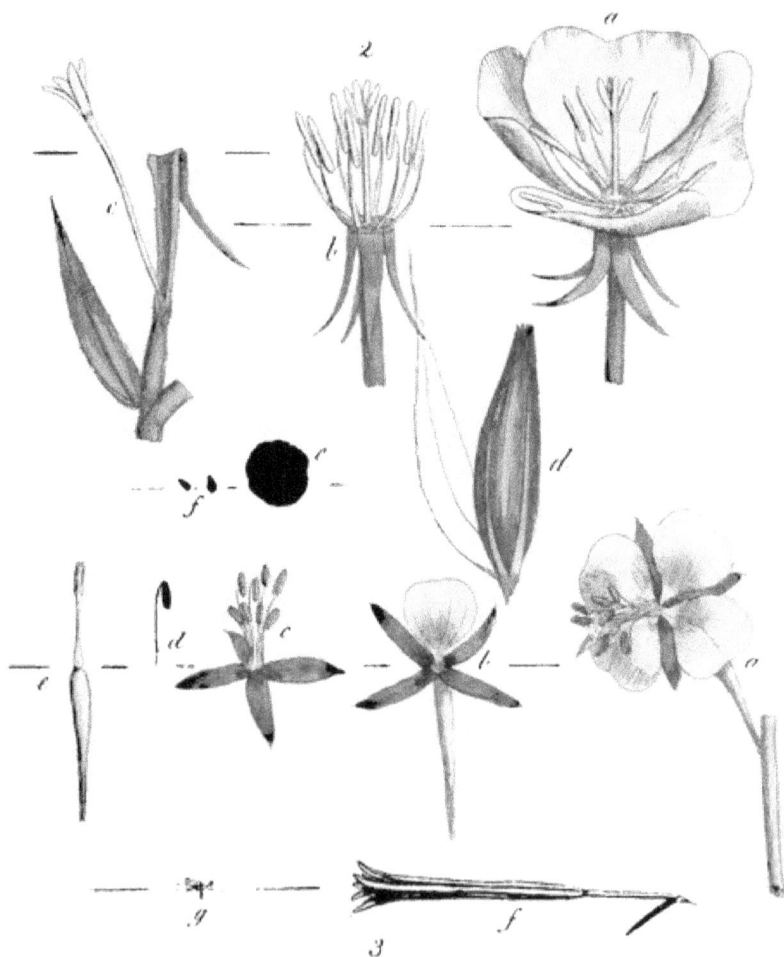

PLATE XV. LETTER XIX.

HEPTANDRIA.

Fig. 1. Æſculus Hippocaſtanum. *Horſe Cheſnut.*

a The corolla of five petals, and the ſeven ſtamens, with bending filaments.
b The one-leafed calyx, ſwelling at the baſe, and divided at top into five ſegments.
c The young capſule terminated by the ſtyle.
d A ſingle ſtamen.

OCTANDRIA.

Fig. 2. Oenothera biennis. *Tree Primroſe.*

a A flower, ſhowing the four-parted calyx, and the corolla of four obcordate petals.
b The eight ſtamens, and the piſtil in the middle, with the deflected calyx.
c The piſtil, with the filiform ſtyle, and the quadrifid ſtigma.
d The capſule.
e A tranſverſe ſection of the capſule, ſhowing the four cells.
f The ſeeds.

Fig. 3. Epilobium anguſtifolium. *French Willow.*

a　The flower.

b　The four-leaved calyx.

c　The ſtamens, four longer and four ſhorter.

d　A ſingle ſtamen.

e　The piſtil.

f　The capſule.

g　A ſeed crowned with down.

PLATE XVI. LETTER XIX.

ENNEANDRIA HEXAGYNIA.

Fig. 1. Butomus umbellatus. *Flowering Ruſh.*

a The flower of ſix petals.
b The nine ſtamens.
c The ſix capſules.

DECANDRIA MONOGYNIA.

Fig. 2. Dictamnus albus. *Fraxinella.*

a The flower, with a corolla of five ſpread-ing petals.

b The five-leaved calyx, with the capſules.

c A ſingle filament, with its glandules.

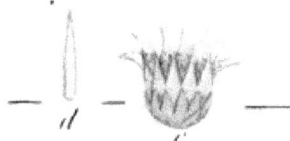

PLATE XVII. LETTER XX.

DODECANDRIA DODECAGYNIA.

Sempervivum tectorum. *Common Houseleek.*

a The flower-stem, with a reflexed range of flowers.

b A flower in front, showing the corolla of twelve petals.

c The calyx, with the capsules, after the flower is past.

d A single capsule.

e The twelve stamens and twelve styles, separated from the flower.

f A single pistil, exhibiting the germ, style, and anther.

g Two stamens.

D

PLATE XVIII. LETTER XXI.

ICOSANDRIA.

Fig. 1. Myrtus communis. *Common Myrtle.*

a The corolla.

b The fruit or berry.

c A fingle flower without the corolla, fhowing the ftamens proceeding from the
calyx.

Fig. 2. Pyrus Cydonia. *The Quince.*

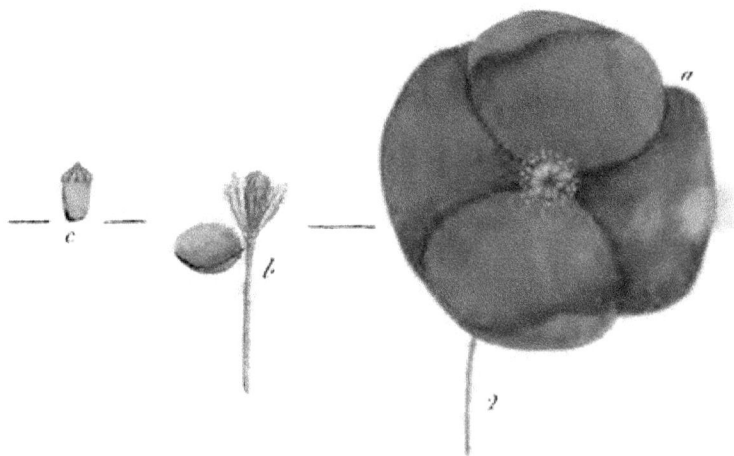

PLATE XIX. LETTER XXI.

POLYANDRIA.

Fig. 1. Caltha paluſtris. *Marſh Marigold.*

a A flower, ſhowing the corolla of five petals, the many ſtamens ſhorter than the corolla, &c.

b Another flower, ſhowing that it has no calyx.

c The capſules, after the flower is paſt.

Fig. 2. Papaver Rhoeas. *Corn Poppy.*

a The corolla of four large roundiſh petals.

b The numerous ſtamens proceeding from the receptacle.

c The capſule crowned with its ſtigma.

Obſ. Fig. 1. is an inſtance of the order Polygynia. Fig. 2. of the order Monogynia.

D 3

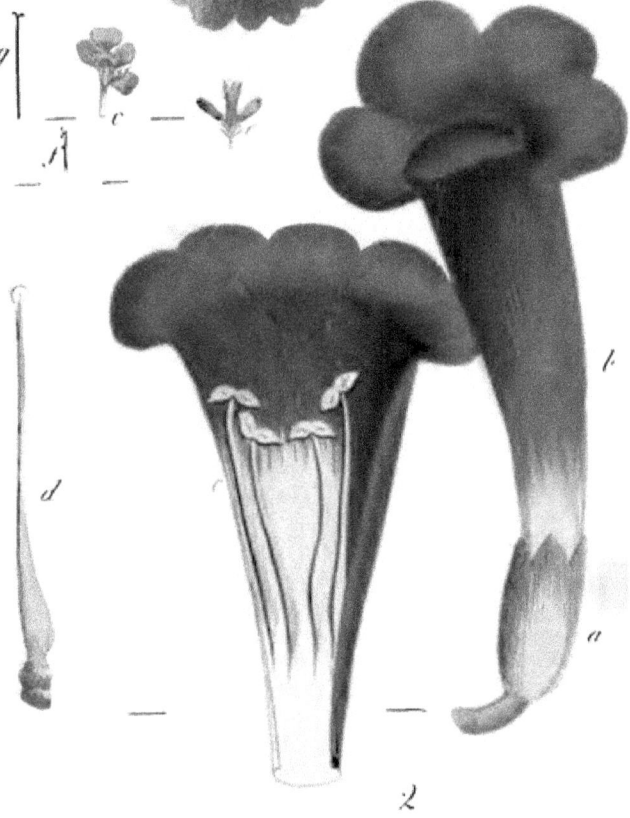

PLATE XX. LETTER XXII.

DIDYNAMIA GYMNOSPERMIA.

Fig. 1. Glechoma hederacea. *Ground Ivy.*

a The kidney-fhaped leaves.

b The ringent flowers.

c A flower opened, to fhow the fituation of the ftamens.

d A flower exhibiting the cruciform appearance of the anthers.

e The calyxes.

f A fingle filament.

g The piftil.

DIDYNAMIA ANGIOSPERMIA.

Fig. 2. Bignonia radicans. *Trumpet Flower.*

a The calyx.

b The corolla.

c The corolla difplayed, to fhow the fituation of the ftamens.

d The piftil.

Obf. The claffical character is clearly fhown at Fig. 2. *c.*

This clafs was farther illuftrated in Plate IV.

D 4

PLATE XXI. LETTER XXIII.

TETRADYNAMIA.

Sifymbrium Nafturtium.   *Water Crefs.*

a a The pinnated leaves.

b   The odd lobe ending blunt.

c   The corymb of flowers.

d   A fingle four-petalled cruciform flower.

e   A fingle petal.

f   The calyx.

g   The calyx, with the ftamens.

h   A fingle ftamen.

i   The filique.

Compare *Plate* XIII.   *See alfo Plate* II.

Plate XXII

PLATE XXII. LETTER XXIV.

MONADELPHIA.

Fig. 1. Althæa officinalis. *Marfh Mallow.*

a   The flower, fhowing the five petals united
at bottom, obcordate or inverfely heart-
fhaped, and flightly emarginated or
end-nicked.   In the centre is the
column of ftamens, with the piftils in
the middle of them.

b   The column of ftamens and piftils removed from the corolla, and fhowing
the rudiment of the fruit underneath.

c   The piftil feparate.

d   The calyx, exhibiting the nine divifions
of the outer calyx, which is one of the
principal generic characters.

Fig. 2. Malva fylveftris. *Common Mallow.*

a   The flower as before.   The petals narrow, heart-fhaped, and much more
deeply end-nicked.

b c The column of ftamens, and piftil feparated.

d   The fruit, with the double calyx; the
outer very narrow, the clefts of the

6

inner broad and large: there are five
of thefe, and three diftinct leaves in the
other; but all of them could not be
reprefented. The fruit flat, with many
feeds in a ring, each covered with its
aril, or loofe coat.

Fig. 3. Geranium zonale. *Horfe-fhoe
Cranefbill.*

a The flower, fhowing the corolla of five
unequal petals, with the column of
ftamens, very flightly connected at
bottom, and of unequal lengths.

b The calyx, with the column of ftamens.
Both thefe figures fhow the ftyle ftand-
ing up above the ftamens, and terminated by five ftigmas.

c The fruit, with the permanent ftyle and
ftigmas; fhowing the beaked form of
it, and the five feeds in their arils, each
terminated by a tail, and feparating
from the beak. *a b c* fhow that the
calyx is fingle and five-leaved.

N. B. Thefe figures ferve to explain the clafs
Monadelphia: and two of the orders,
Decandria, Fig. 3, and *Polyandria*,
Fig. 1, 2.

Drawn & Engraved by J. P. Smith.

PLATE XXIII. LETTER XXV.

DIADELPHIA DECANDRIA.

Lathyrus latifolius. *Everlasting Pea.*

Fig. 1.   A bunch of flowers, in their natural fize and fituation.

Fig. 2.   The banner.

Fig. 3.   One of the wings.

Fig. 4.   The keel.

Fig. 5.   The ftamens and piftil in their natural fituation.

Fig. 6.   The ftamens, fhowing the fimple filament feparate from the compound one.

Fig. 7.   The piftil.

See Plate III.

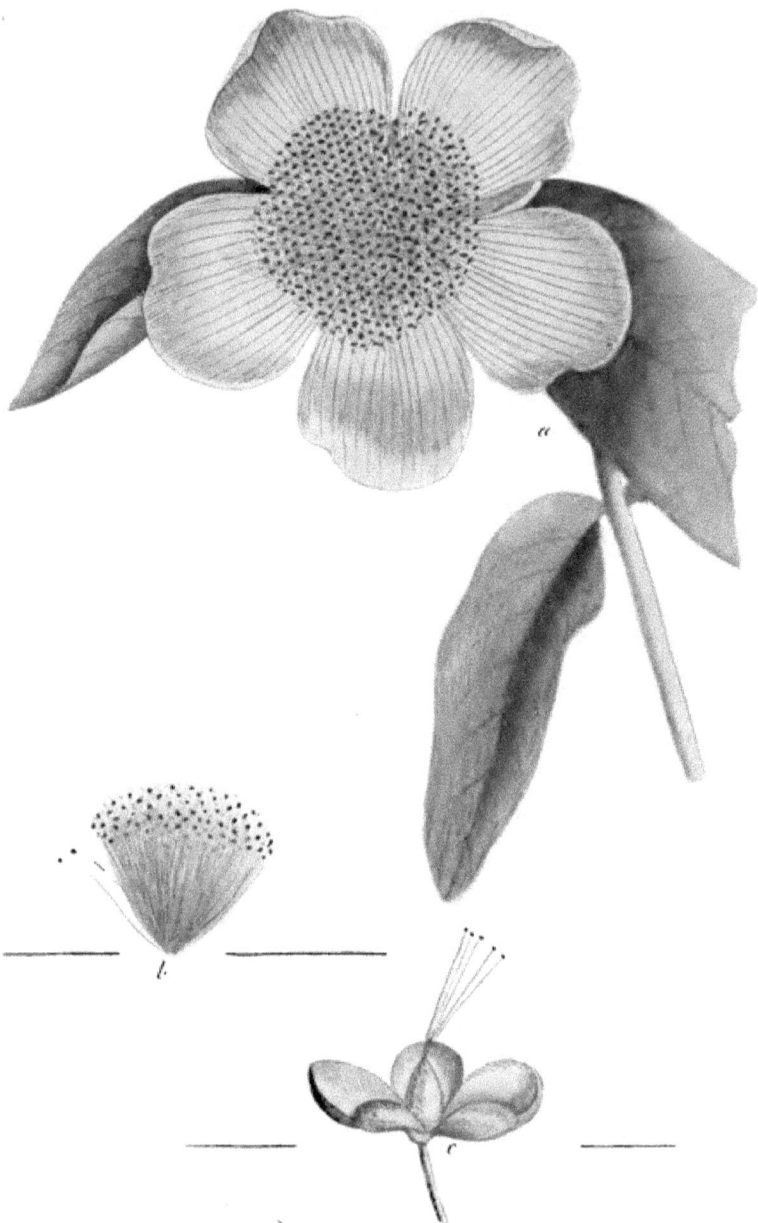

on it. engraved by E. P. N. Abe.

PLATE XXIV. LETTER XXV.

POLYADELPHIA.

Hypericum Afcyron. *Garden Tutfan.*

a The flower, with a corolla of five petals and the numerous ftamens in the middle.

b A fingle pencil or parcel of ftamens.

c The permanent five-parted calyx, including the germ terminated by five piftils.

b Explains the characters of the clafs and order—Polyadelphia Polyandria.

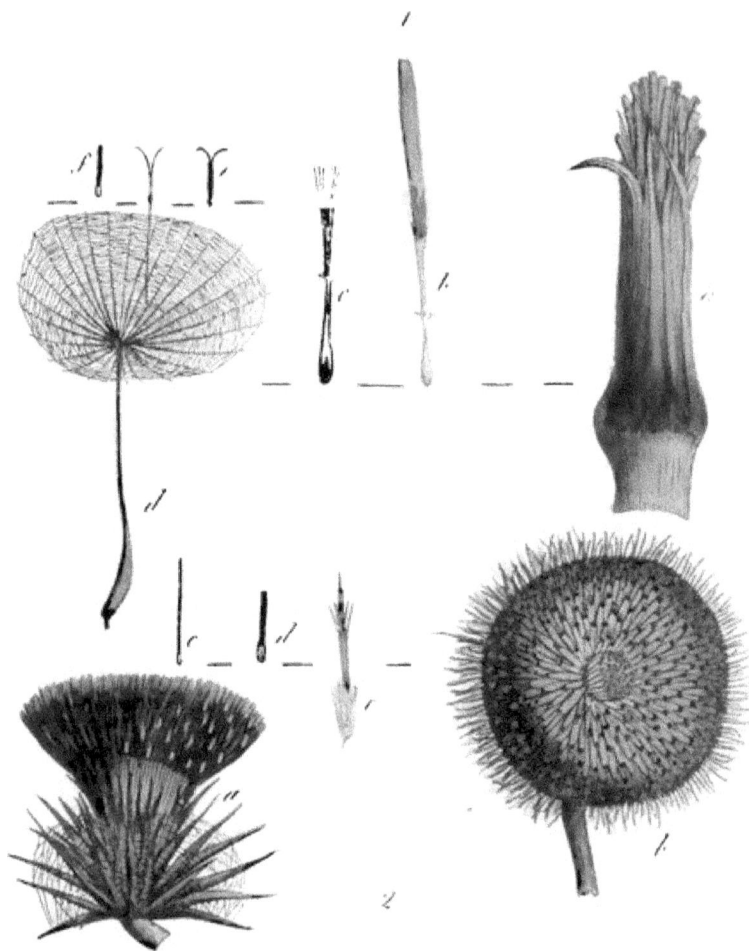

PLATE XXV. LETTER XXVI.

SYNGENESIA POLYGAMIA ÆQUALIS.

Fig. 1. Tragopogon porrifolium. *Salfafy.*

a A flower clofed, fhowing the fimple calyx.

b A fingle ligulate flofcule.

c A flofcule, deprived of the corolla.

d A feed, with the feathered ftipitate down.

e The cylinder of anthers, with the piftil perforating it, terminated by the two revolute ftigmas.

f The cylinder of anthers alone.

Fig. 2. Carduus nutans. *Mufk Thiftle.*

a The compound flower, fhowing the calyx all imbricate with thorny fcales.

b A front view of the whole compound flower, compofed wholly of tubulous florets.

c A fingle flofcule or floret.

d The cylinder of anthers.

e The piftil.

E

Fig. 3. Eupatorium cannabinum. *Common
Hemp Agrimony.*

a A bunch of flowers.
b A single flower.
c A single bunch of flowers.
d The down.

Obf. Thefe three figures explain the three
fections of this order. 1. Containing compound flowers with ligulate florets only. 2. The capi-
tate or headed flowers, with tubulous florets only. 3. The dif-
coid, or naked difcous flowers,
with tubulous florets, but not in a
head.

Plate XXVI.

Drawn & Engraved by F.P. Nodder

Publish'd 1 May 1798 as the Act directs by B White & Son.

PLATE XXVI. LETTER XXVI.

SYNGENESIA POLYGAMIA SUPERFLUA.

Doronicum pardalianches. *Common Leopard's*
Bane.

a   The compound radiated flower, confifting
of regular tubulous flofcules in the
difk, and irregular ligulate flofcules in
the ray.

b   The under part of the flower, fhowing
the double row of fcales to the calyx.

c   One of the femi-florets, or ligulate flofcules, taken from the ray, to fhow
that the feed is naked, or deftitute of
down.

d   A floret from the difk, the feed of which
is crowned with a fimple down.

e   A fection of the difk, in order to exhibit
the naked receptacle.

E 2

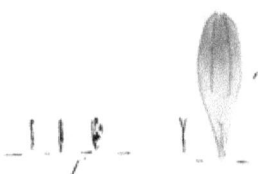

PLATE XXVII. LETTER XXVI.

SYNGEN. POLYG. FRUSTRANEA
and NECESSARIA.

Fig. 1. Centaurea montana. *Mountain Blue Bottle.*

a The compound flower, showing the neutral or barren florets on the outside, longer than the fertile ones in the middle, and the ciliated scales of the calyx.

b A barren floret.

c A fertile floret, with some of the bristles at the base.

d The same, divested of the corolla.

e The piftil.

N. B. This serves to explain the order Polygamia Frustranea in the class Syngenesia.

Fig. 2. Calendula officinalis. *Garden Marigold.*

a The compound radiated flower.

b The calyx, with the seeds in the ray only, bending inwards after the florets are decayed.

E 3

c The boat-fhaped muricated feed, without down.

d A barren feed, from one of the central flowers.

e A feitile flofcule from the ray.

f A barren flofcule from the difk.

N. B. This ferves to explain the order Poly-gamia Neceffaria in the clafs Syn-genefia.

PLATE XXVIII. LETTER XXVI.

SYNGEN. POLYG. SEGREGATA.

Echinops fphærocephalus. *Globe Thiſtle.*

a   The entire compound flower, confifting of tubular florets, feparated by their proper perianths; which determines this plant to be of the fegregate order in the clafs Syngenefia.

b   A finuated leaf, the jags ending in fpines.

c   A fingle flofcule in its calyx.

d   A flofcule taken out of the calyx, with the ftyle feparate.

e   A fingle fubulate leaflet of the calyx, in three different views.

PLATE XXIX. LETTER XXVI.

SYNGENESIA MONOGAMIA.

Viola odorata. *Sweet Violet.*

a   The calyx of five leaves.

b   The corolla of five irregular petals.

c   The horn-shaped nectary.

d   A flower opened, to show the stamens with the five connected anthers.

e   The stamens within the calyx.

f   A single stamen.

g   The pistil.

h h h   The heart-shaped leaves.

i i   The young leaves, involuted, rolled inwards, or rather upwards.

k k k   The scape, with the double bracte on the middle of it.

l   One of the stolones, or runners, putting forth roots.

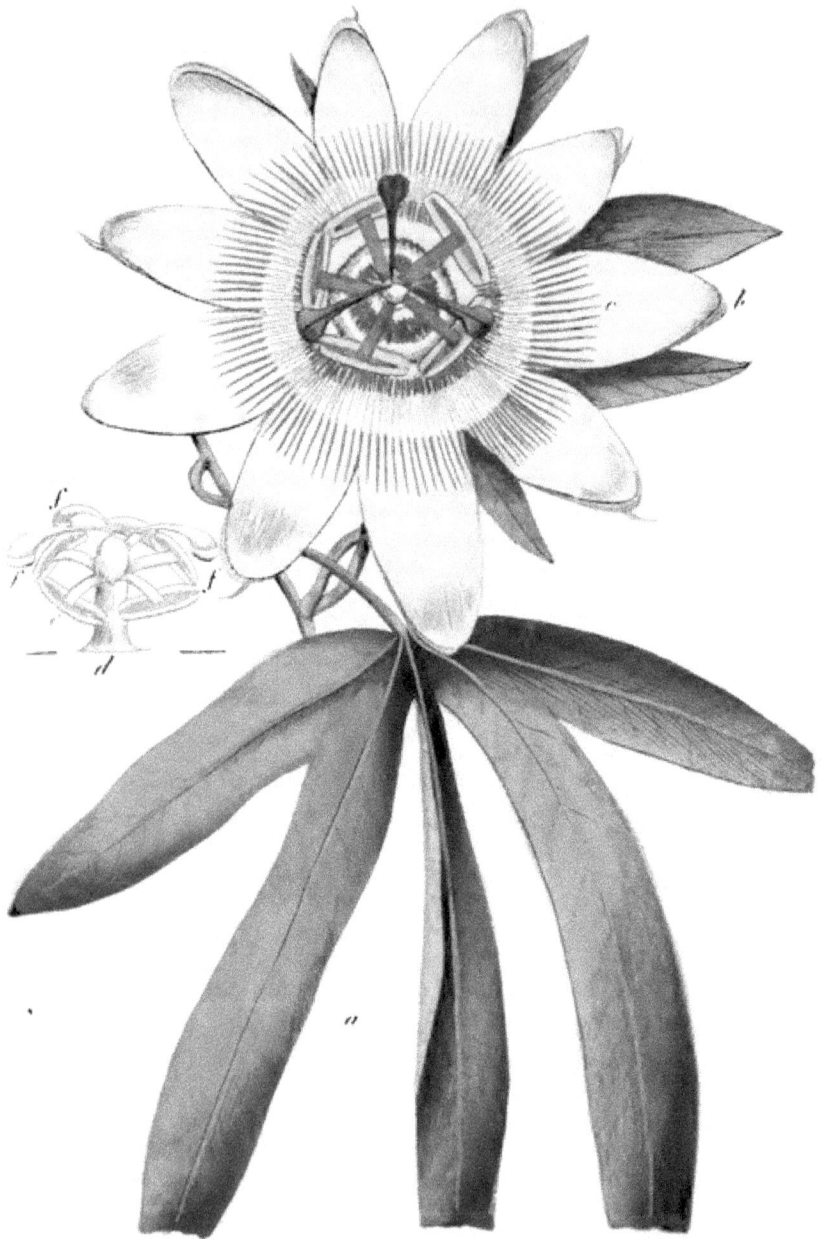

Drawn & engraved by J. P. Nuttle.

PLATE XXX. LETTER XXVII.

GYNANDRIA.

Paffiflora cærulea. *Blue Paffion Flower.*

a The palmated leaf.

b The corolla and calyx, each of five leaves, and having the fame appearance in front.

c The radiate crown, which is the nectary.

d The piftil and five ftamens.

e The anthers terminating the filaments, which fpring from the bottom of the germ, where it meets the pedicle, upon which it ftands.

f f f The three ftigmas arifing from the germ.

Pl. XXXI

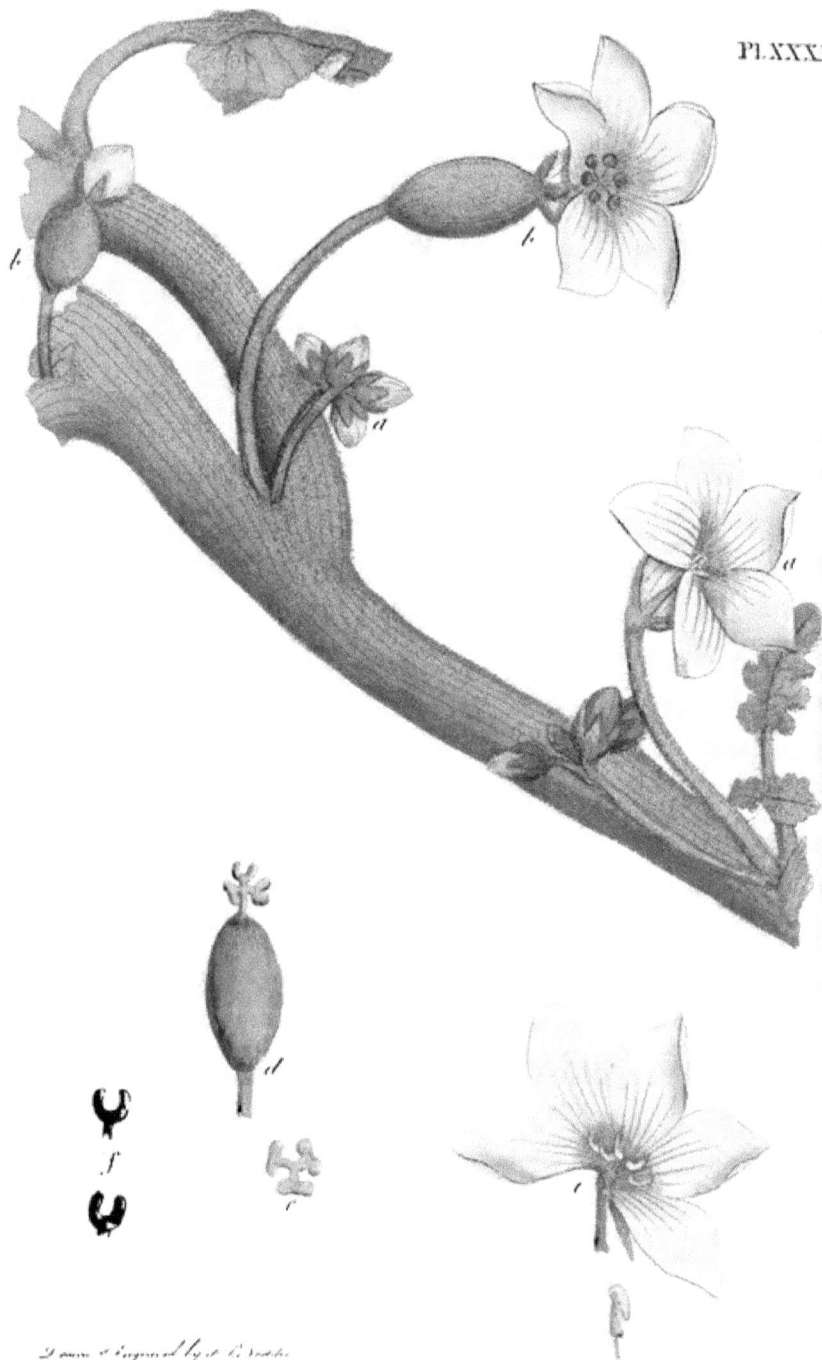

PLATE XXXI. LETTER XXVIII.

MONOECIA.

Momordica Elaterium. *Spirting Cucumber.*

a a   The male or ftaminiferous flowers.

b b   The female or piftilliferous flowers, with
the large germ below the receptacle.

c   The male flower, fhowing the three filaments, with double anthers on two
of them, and a fimple anther on the
third.

d   The germ, furmounted with the ftyle,
divided into three parts, each part
fuftaining an oblong gibbous ftigma.

e   The divided part of the ftyle, with the
ftigmas.

f   Two different views of a fingle ftigma.

PLATE XXXII. LETTER XXIX.

DIOECIA.

Cannabis fativa.   *Hemp.*

Fig. 1.   *Female Hemp.*

a   A fingle female flower.
b   The feed included within the calyx.

Fig. 2.   *Male Hemp.*

a   Male flowers feparate.

Pl. XXXIII.

PLATE XXXIII. LETTER XXX.

POLYGAMIA MONOECIA.

Acer campeftre.　*Common Maple.*

a a 　The lobed leaves.

b b 　Bunches of flowers. — *c* Perfeſt. — *d* Male, with ſtamens only.

e 　A ſingle perfeſt flower.

f 　A petal.

g 　A perfeſt flower divefted of the corolla and calyx.

h 　A ſingle ſtamen.

i 　The piſtil, with the two revolute ſtigmas, and the rudiment of the two capſules, terminating in a wing.

k 　A male, or ſtaminiferous flower, and a ſingle petal of it.

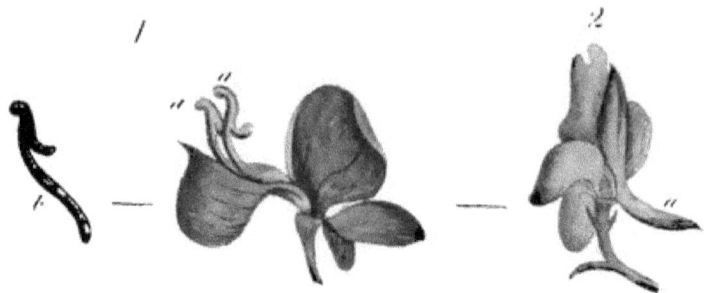

PLATE XXXIV. LETTER XXXI.

NECTARIES.

Fig. 1.   Aconitum Napellus.   *Blue Monk's Hood.*

a a   The two recurved pedunculated nectaries.

b   A fingle nectary, taken out of the flower.

Fig. 2.   Delphinium Ajacis.   *Garden Lark-fpur.*

a   The nectary, continued backward in form of a horn or fpur.

Fig. 3.   Parnaffia paluftris.

a   A flower, with the nectareous fcales at the bafe of the ftamens.

b   The five heart-fhaped nectaries, terminating in hairs, with a little ball on the top of each, and placed between the ftamens.

Fig. 4.   A petal of the Ranunculus, fhowing the honied gland juft above the bafe, on the infide at *a a.*

F 2

Fig. 5.   Iris or Flag. The nectary, in form
of a villous line, along the middle of
one of the reflex petals.

Fig. 6.   Fritillaria Imperialis.   *Crown Im-
perial.*

a   An excavation at the bafe of the petal,
which is the nectary.

Fig. 7.   Afphodelus luteus.   *Yellow Afphodel.*

a   The flower, fhowing the fix ftamens, each
fitting on its valve, and the fix valves
forming an arch over the germ.

b   A fingle filament on its fcale, which is
inferted into the bafe of the petal.

Fig. 8.   Helleborus foetidus.   *Stinking* **Black-**
Hellebore.

a   The tubular nectaries placed in a ring at
the bafe of the ftamens.

b   A fingle nectary.

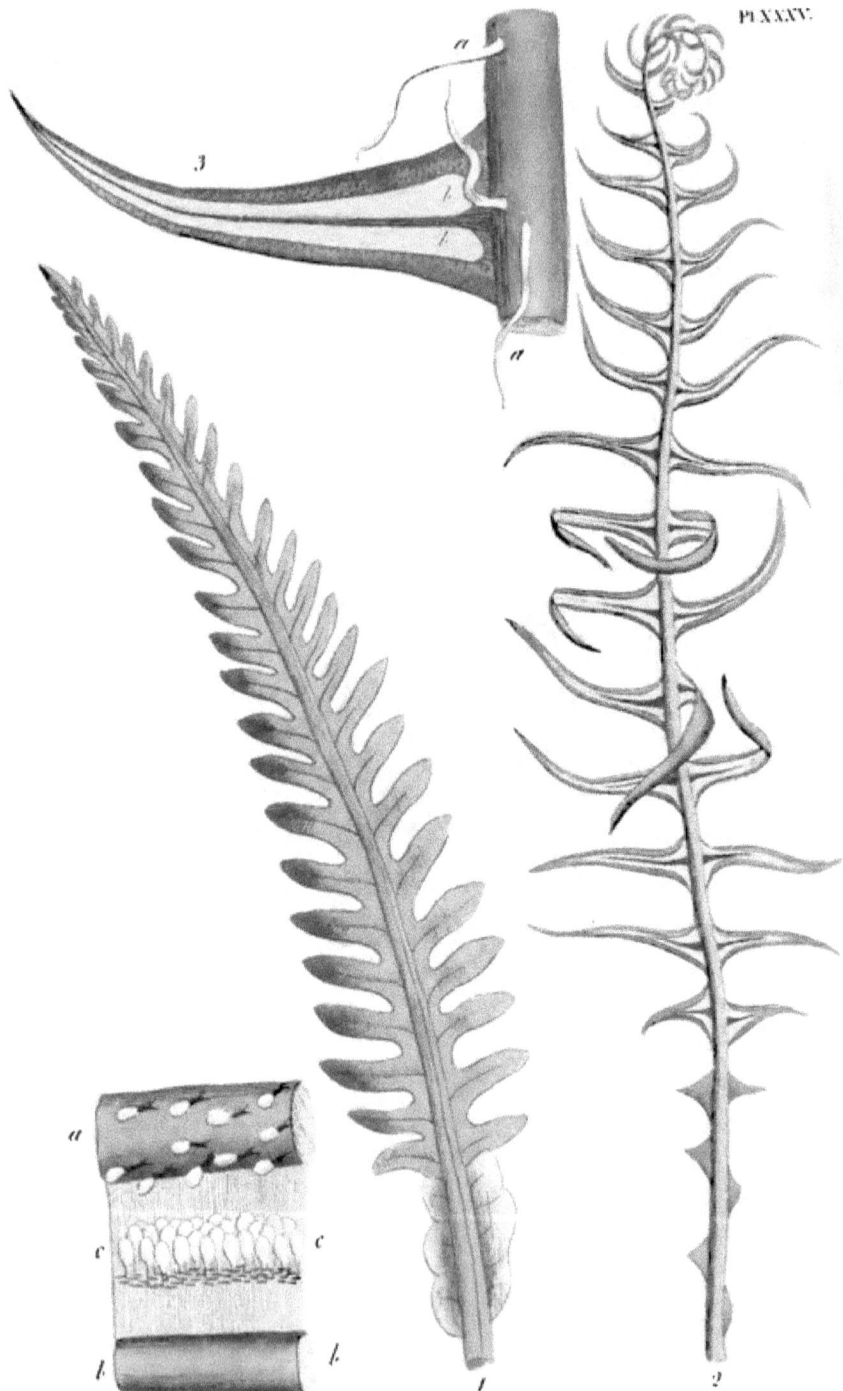

Pl. XXXV.

PLATE XXXV. LETTER XXXII.

CRYPTOGAMIA FILICES. Ferns.

Ofmunda Spicant. *Rough Spleenwort.*

Fig. 1.   The barren frond.

Fig. 2.   The fertile frond.

Fig. 3.   A fingle pinna magnified, with the fcales at *a a*; and covers of the capfules at *b b*.

Fig. 4.   A part of the pinna more magnified, with the anthers on the rib at *a*, and the membrane rolled back at *b b*, to exhibit the rudiments of the feed veffels at *c c*.

PLATE XXXVI. LETTER XXXII.

CRYPTOGAMIA MUSCI. Moffes.

Bryum pyriforme. *Pear Bryum.*

Fig. 1.   The mofs of its natural fize.

Fig. 2.   The anthers yet entire.

Fig. 3.   The female flower, while it is yet inclofed within the inmoft leaves.

Fig. 4.   The fame feparated, with the appendages, viz. *a a* the adductors. *b b* the cylindrical jointed threads.

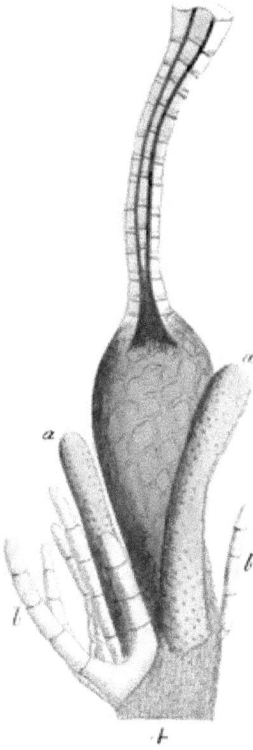

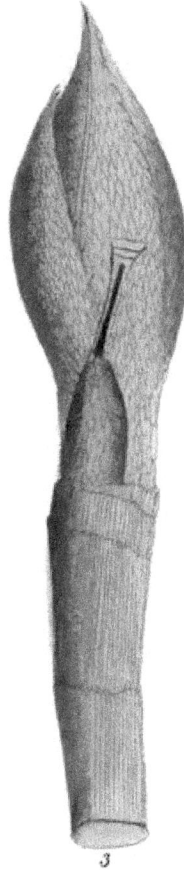

Pl XXXVII.

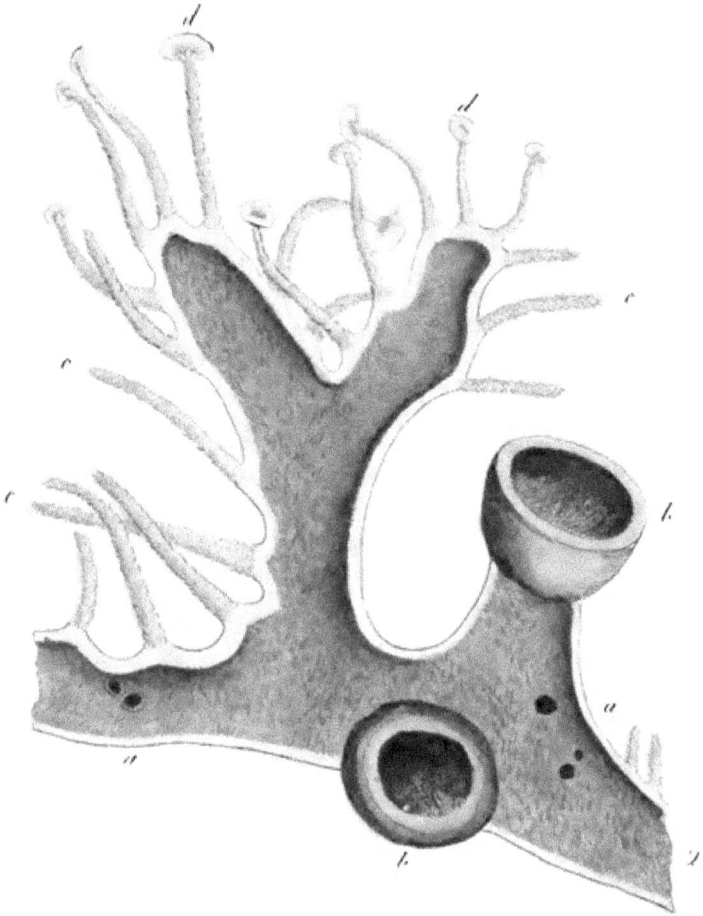

PLATE XXXVII. LETTER XXXII.

CRYPTOGAMIA ALGÆ.

Lichen ciliaris. *Ciliated Liverwort.*

Fig. 1. The plant of its natural fize.

Fig. 2. The fame magnified.

a a The male or barren flowers.

b b The females in a ftate of ripenefs.

c c The rooting hairs.

d d The hairs, or ciliæ, growing on the extremities.

Fig. 3. The feeds magnified.

PLATE XXXVIII. LETTER XXXII.

CRYPTOGAMIA FUNGI. Fungufes.

Agaricus Dillen. giff. p. 185.

Fig. 1.   Plants of different ages, and of their natural fize.

a   Is the Fungus in its perfect or adult ftate.

b   The fame in its middle ftate.

c   Small plants juft rifing.

Fig. 2.   A parcel of knotted threads from the fungus marked *b*, fuppofed to be the ftamens.

Fig. 3.   A fection of the cap *(a)* and la- mella *(b)* of the fame fmall fungus magnified.

Fig. 4.   The ripe feeds of this fungus much magnified.

Obf. Thefe four plates are copied from Hedwig's Theoria, as it would have anfwered little purpofe to figure fuch minute plants of their natural fize only.

THE END.

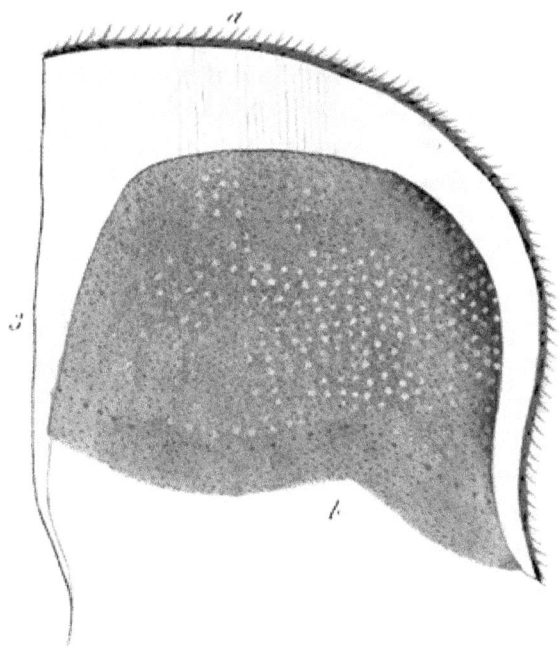

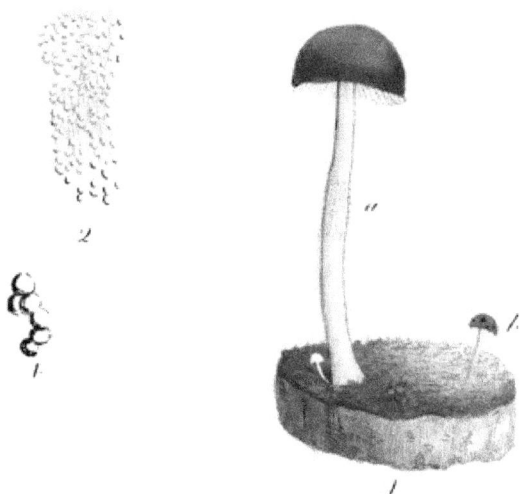

BOOKS

Lately *Published by* JOHN WHITE.

1. Letters on the Elements of BOTANY. Addressed to a Lady, by the celebrated J. J. ROUSSEAU. Translated into English, with Notes, and twenty-four additional Letters, fully explaining the System of Linnæus. By THOMAS MARTYN, B.D. F.R. and L.S.S. Professor of Botany in the University of Cambridge. The Fourth Edition. Price Seven Shillings, in Boards.

2. The LANGUAGE of BOTANY : being a Dictionary of the Terms made use of in that Science, principally by Linnæus : with familiar Explanations, and an Attempt to establish Significant English Terms. The whole interspersed with Critical Remarks. By THOMAS MARTYN, B.D. F.R. and L.S.S. &c. Price Five Shillings, in Boards.

3. FLORA DIÆTETICA; or, History of Esculent Plants, both domestic and foreign : in which they are accurately described, and reduced to their Linnæan Generic and Specific Names; with their English Names annexed, and ranged under eleven General Heads, viz. 1. Esculent Roots; 2. Shoots, Stalks, &c; 3. Leaves; 4. Flowers; 5. Berries; 6. Stone Fruit; 7. Apples; 8. Legumens; 9. Grain; 10. Nuts; 11. Fungus's; and a particular Account of the Manner of using them; their native Places of Growth; their several Varieties and Physical Properties: together with whatever is otherwise curious or remarkable in each Species. By CHARLES BRYANT. Price Six Shillings, in Boards.

4. Elements of CONCHOLOGY : or, an Introduction to the Knowledge of Shells. With seven Plates, containing Figures of every Genus of Shells. By EMANUEL MENDES DA COSTA. Price Seven Shillings and Sixpence; or, the Plates beautifully coloured, Fifteen Shillings, in Boards.

5. The NATURALIST's JOURNAL, upon the Plan of Mr. Stillingfleet, for keeping a daily Register of Observations on the Weather, Plants, Birds, Insects, &c. By the Honourable DAINES BARRINGTON: a new Edition, neatly engraved, and printed on a fine Writing Paper. Price Five Shillings, sewed in Marble Paper.

6. The Natural History of many curious and uncommon ZOOPHYTES, collected from various Parts of the Globe. By the late JOHN ELLIS, Esq. F. R. S. Systematically arranged and described by the late DANIEL SOLANDER, M. D. F. R. S. &c. with sixty two very elegant Plates. Price One Pound Sixteen Shillings, in Boards.

7. New Illustrations of ZOOLOGY, intended as a Supplement to Edwards's Natural History of Birds. By PETER BROWN. With fifty Plates of new, curious, and non-descript Birds, Quadrupeds, &c. most beautifully coloured. Price Three Guineas, half bound.

8. Handsomely printed on a fine Paper, in One Volume Quarto, ornamented with elegant Engravings, The Natural History and Antiquities of SELBORNE, in the County of Southampton; in a Series of Letters to THOMAS PENNANT, Esq. and the Honourable DAINES BARRINGTON. By the Reverend GILBERT WHITE, M. A. late Fellow of Oriel College in Oxford. Price one Guinea, in Boards.

9. The LIFE of LINNÆUS, with a List of his Works, and a Biographical Sketch of the Life of his SON. Translated from the Original German, of D. H STOEVER, Ph. D. by JOSEPH TRAPP, A. M. Dedicated to the Linnæan Society; with a Portrait of Linnæus, by Heath. Price One Guinea, in Boards.

10. The same on Large Paper. Price One Pound Seven Shillings.

www.ingramcontent.com/pod-product-compliance
Lightning Source LLC
Chambersburg PA
CBHW021812190326
41518CB00007B/561